Introc
Gambling Theory

Know the Odds!

ALSO BY WILLIAM R. PARKS

Paperback editions

Beginning Algebra

Sets and Numbers

Algebraic Expressions

Linear Equations and Graphs

Properties of the Real Number System

Computer Mathematics

Sets and Flowcharts

Computer Number Bases

Introduction to Logic

Boolean Algebra and Switching Circuits

Statistics

Introduction to Statistics

Introduction to Gambling Theory

Introduction to Gambling Theory

Know the Odds!

William R. Parks, B.S., M.S., Ed.M.

Hershey Books

www.wrparks.com

eBook ISBN: 978-0-88493-029-7

Paperback edition Library of Congress Control Number: 2015919819

INTRODUCTION

This study covers the fundamental concepts of gambling theory in a simple and easy to understand format. We utilize a modified form of programmed instruction (PI) in this presentation. Small bits of information are presented in each frame before advancing to the next frame. Exercises are listed after several frames, followed by an answer key.

We also use example problems with an immediate presentation of the solution given right after the problem. For example, if we propose a problem such as how many times a head will show up when you toss a coin 10 times, we immediately propose a solution for predicting the outcome of 10 tosses.

The immediate answer feedback approach to a problem is important when learning new concepts. The exercises serve the same purpose. Answers are given right after each set of exercises.

Before discussing specific rules and theories about gambling itself, we first introduce the reader to important concepts of probability theory. Probability theory is the basis for gambling theory. You cannot learn the rules of gambling without first having some knowledge of probability theory.

Probability theory and gambling theory go hand and hand. The mathematics in probability theory is used to explain happenings in gambling.

INTRODUCTION TO GAMBLING THEORY

1. The basis for understanding gambling theory is probability theory. Few gamblers take time to acquire this knowledge about probability.

DEFINITION:
Probability theory can be defined as the study of determining the likelihood that some future event will take place or that a series of events will take place. Gamblers are happiest when events take place in their favor when they gamble.

Probability theory studies ways to predict the likelihood that certain events will take place at a gambling table within a calculated percent such as 25% or 50% or 75% of the time. These same three probabilities can also be expressed as fractions:

$\frac{1}{4}$, $\frac{1}{2}$ and $\frac{3}{4}$

¼ means the desired event might take place one out of four times or 25% of the time.

Probability is concerned with the study of random events. At first glance it seems unlikely that a scientific study or a theory could be developed about random events. For example, we cannot tell in advance or with certainty that a seven will turn up in a single roll of two dice.

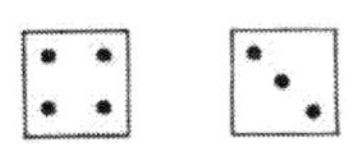

If the dice were loaded and this caused a seven to turn up more often, we still would not know for certain if a seven will turn up on the next roll.

We can make some sensible predictions if we become familiar with the laws of probability. We will learn that the probability of rolling a seven with a fair die is:

$\frac{1}{6}$ or 16.67%.

Blaise Pascal (1623 – 1662), a famous mathematician and philosopher, joined other mathematicians in developing a general theory of probability in the seventeenth century. It was applied to gambling theory and rolling dice. Later it became important and useful in making predictions and explaining uncertainties in economics and actuarial science.

Mathematicians after Pascal built on his ideas and came up with greater knowledge about probability and gambling theory.

\- - - - - - - - - - - - -

2. There are two ways to study probability:

(a) By observation called the empirical method.

Data is collected over a period of time and it is used to predict the future.

(b) By "a priori" determination, called classical probability.

The words "a priori" mean that a prediction of future events is made without prior observation of past events. Reasoning, mathematics and probability theory are used to make predictions rather than base predictions on observations or past experiences.

3. In the empirical method we predict that some future event might likely take place based on some past observations.

EXAMPLE:
If we tossed a fair coin 50 times, we might observe that 25 heads and 25 tails appeared out of 50 tosses.

QUESTION:
What would you predict for the next 50 tosses based on the above past observations?

ANSWER:
You might predict that the probability would stay close to 50% heads and 50% tails, or 25 heads and 25 tails.

EXPERIMENT:
Toss a fair coin 20 times and record the number of heads and number of tails. Were your results close to 50% heads and 50% tails? If yes, the empirical method worked.

\- - - - - - - - - - - - - -

4. In classical "a priori" probability we predict that some future events can take place on the basis of pure mathematical calculations rather than making predictions based on past observations.

Deductive logic and mathematical calculations are used to predict the likelihood of future events. One can calculate the likelihood that an event will take place by learning to apply specific formulas found in "a priori" probability theory.

.

EXAMPLE:
We can calculate the probability of tossing a head on a single toss of a fair coin without any record of past observations. For example we know for a fact that a coin has two sides: a head and a tail. We desire a head and we know there are two possibilities. This "a priori" determination can be expressed by a ratio formula. The number of desired events is put in the numerator and the number of total possible events is put in the denominator as follows:

$$\text{Probability} = \frac{\text{the number of desired outcomes}}{\text{total number of possible outcomes}}$$
$$= \frac{\text{one}}{\text{two}} = \frac{1}{2}$$

The fraction one-half means one out of two tosses should yield a head. Probability is expressed mathematically as a fraction. In this case it is expressed as one over two. Since there are two faces on a coin, tail and head, there are two possible outcomes. The number two is placed in the denominator. Only one out of two faces has a head on it. Therefore, the number one is put in the numerator. The probability of tossing a head is one out of two or one-half which can also be expressed as .5 or a 50% likehood of tossing a head.

$$\frac{\text{one head}}{\text{two sides}} = \frac{1}{2} = .5 = 50\%$$

This probability does not guarantee that every two throws of the coin will yield a head. It is in the long run of many tosses that we expect a ratio of one head out of every two tosses. If you toss a coin enough times, the occurence of a head 50% of the time will gradually become evident.

To summarize – The probability of an event taking place such as in coin tossing is equal to the number of times that an event can happen divided by the total number of possible outcomes (with the assumption that a fair coin is used).

In actual practice what is observed taking place may differ from the predicted classical a priori determination. What we observe is called empirical probability, also known as relative frequency, or experimental probability. The ratio is the number of outcomes a specified event occurs divided by the total number of possible trials. In a real world experiment. The empirical probability (E.P) formula is as follows for coin tossing a head:

$$E.\ P. = \frac{\text{number of observed heads}}{\text{the total number of outcomes}}$$

If we tossed an unfair weighted coin 100 times and got 40 heads, then the E.P. for heads in this experiment would be:

$$\frac{40}{100} = 40\%\ \text{likelihood of getting heads with this unfair coin.}$$

In casino gambling "a priori" determined answers can help to pick a winner but there is no guarantee you will do so and win each time that you bet because actual events are unpredictable and "a priori" formulas only work in the long run after many trials. Also, keep in mind that casinos know the formulas and adjust the game rules to give the house a considerable advantage to win more often than most players. Gamblers lose more often than they win.

5. EXERCISE 1

Answer all questions before checking the answer key in Frame 6 on the next page.

1. A basis for studying gambling theory is a knowledge of ____________________theory.

2. The theory of probability can be defined as a study of ways to determine the likelihood that some _________ _________ will take place.

3. In your own words describe the two approaches that are used to study probability:

(a) Empirical method: ____________________________

____

____

(b) Classical method____________________________

____

____

4. Sam Smith tossed a coin 10 times and got 6 heads and 4 tails. (a) What is the observed outcome of heads in this experiment using the empirical method.

b) How was your answer in (a) different from classical probability using "a priori" determination

6. ANSWERS FOR EXERCISE 1

1. probability

2. future event

3. (a) In empirical probability the likelihood of futre events taking place are based on calculations using data obtained from observing past events.

(b) In “a priori” probability the likelihood of future events is based on calculations in pure mathematics rather than past observations.

4. (a) Empirical probability (E.P.) based on observations:

$$\text{E. P.} = \frac{\text{number of observed heads}}{\text{the total number of outcomes}}$$

$$\text{E. P.} = \frac{6}{10} = \frac{3}{5} = .6 \text{ or } 60\% \text{ heads}$$

(b) $$P(H) = \frac{1}{2} = .5 \text{ or } 50\% \text{ heads}$$

The above statement is read as follows: “The probability of heads is one-half which means it can occur 50% of the time.”

7. The set listing all possible outcomes of an event or in an experiment is called a <u>sample space.</u>

EXAMPLES:

(a) The total number of possible outcomes or sample space when tossing a fair coin is two outcomes: a head and a tail. This is expressed in set notation as: { H, T }. We indicate here two different possible outcomes: a head and a tail.

(b) The total number of possible outcomes (or sample space) when rolling a single die is { 1, 2, 3, 4, 5, 6 }. There are six different possible outcomes.

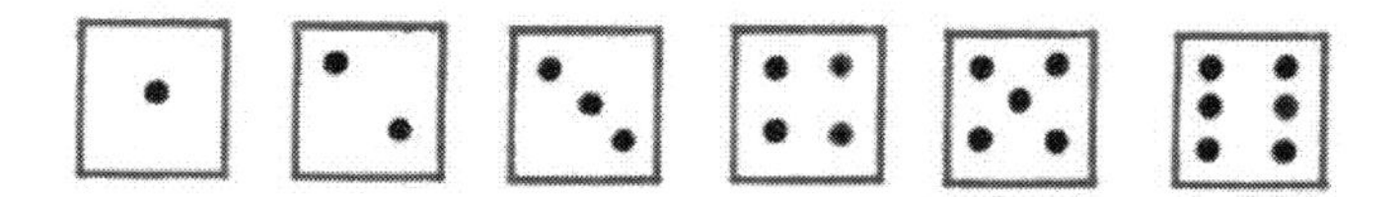

(c) The total number of cards in a standard deck of cards is 52. There are 13 hearts, 13 diamonds, 13 spades and 13 clubs. Each suit has 13 ranks: Ace, 2, 3, 4, 5, 6, 7, 8, 9, 10, jack, queen, king.

In a well shuffled deck of cards, there is a 25% likelihood of selecting a heart at random from a 52 card deck because there are 13 hearts out of 52 cards. We use the probability formula to show how we arrived at 25% as the probability of picking a heart. We designate this as follows:

$$\text{Probability} = \frac{\text{the number of successful outcomes}}{\text{total number of possible outcomes}}$$

$$P\,(H) = \frac{13 \text{ hearts}}{52 \text{ cards}} = \frac{1}{4} = .25 = 25\%$$

8. Other terms often used by gamblers are odds in favor and odds against.

EXAMPLES:
(a) The odds in favor of selecting hearts from a standard 52 card deck is expressed by the following "odds" formula.

Odds are expressed as "chances for" vs "chances against"

The odds or chance of selecting a hearts is 13 : 39 spoken as "the chance of selecting a heart is 13 to 39"

We arrived at this ratio by using the following formula:

Odds = (chances for) : (chances against)

Odds = 13 : 39

In this formula the second term 39 is found by evaluating the following formula:

(Chance against) = (total chances) - (chances for)

= 52 - 13

= 39

(b) The odds in favor of tossing a heads with a fair coin is "one to one" because:

Odds = (chances for) : (chances against)

= 1 : 1

We'll review and cover odds in greater detail in Frame 17.

9. EXERCISE 2

Answer all questions before checking the answers in Frame 10 on the next page.

1. List the sample space or total possible outcomes for the following:
(a) Possible suits in a standard 52 card deck.

(b) Rolling a single die.

(c) Traffic signal lights

(d) Weather in summer

2). Express the probability in three ways: as a fraction, as a decimal and as a percent when selecting a card from the following standard 52 card deck:
(a) Ace

(b) Face Card

(c) Clubs

3. (a) What are the odds or chances of selecting a face card from a standard 52 card deck.

(b) What is the probability of selecting a Face Card?

10. ANSWERS FOR EXERCISE 2

1. (a) { Clubs, Diamonds, Hearts , Spades }

(b) { 1, 2, 3, 4, 5, 6 }

(c) { Green, Red, Yellow, Green Arrow }

(d) { Sunny, Cloudy, Partly Cloudy, Windy, Rain, Hail }

Note: (c) and (d) can have more conditions listed.

2. (a) $P(A) = \frac{\text{4 Aces}}{\text{52 Cards}} = \frac{1}{13} = .0769 = 7.69\%$

(b) $P(F) = \frac{\text{12 Face Cards}}{\text{52 Cards}} = \frac{3}{13} = .2307 = 23.07\%$

(c) $P(C) = \frac{\text{13 Clubs}}{\text{52 Cards}} = \frac{1}{4} = .25 = 25\%$

Note: In 2. (a) & (b) we carried out our answer to three decimal places because they are repeating decimals. We decided that the likelihood of selecting an Ace is 7.69% and of selecting a Face Card is 23.07% which is pretty close to the actual value.

3. (a) Odds of selecting FC = (chances for) : (chances against)
= 12 : 40

The chance or odds of selecting a face card is 12 to 40 or if we simplify and reduce: 3 to 10.

(b) The probability is 23.07%. Same answer as 2.(b).

11. RANGE OF VALUES OF PROBABILITY

Fractions, decimal numbers and percents are used to represent probability. The probability of an event happening is represented by the simple formula we covered in previous sections.

$$\text{Probability (event)} = \frac{\text{the number of successful outcomes}}{\text{total number of possible outcomes}}$$

EXAMPLES:

(a) Probability (rolling an even number on a single toss of one die) $= \frac{\text{three even numbers}}{\text{six possible outcomes}} = \frac{3}{6} = \frac{1}{2} = .5 = 50\%$

In the above problem there are three even numbers { 2. 4. 6 } within the sample space containing six numbers: { 1, 2, 3, 4, 5, 6 }. Therefore, the probability of rolling an even number is one-half or 50%.

The range of values of probability is quantified as a number from 0 to 1 (from 0% to 100%). 0% represents impossibility and 1 or 100% represents absolute certainty that an event will happen. The higher the probability of an event, the more certain that it will happen.

(b) Probability (rolling a number different from 1 to 6 on a single roll of a die) $= \frac{0}{6} = 0 \text{ or } 0\%$

When there is no successful outcome possible, then the probability is zero.

(c) The probability of rolling a number from 1 to 6 with a single roll of a die is $\frac{6}{6} = 1 = 100\%$ certain.

12. The abbreviated formula for probability that a favorable event "A" will occur is often given in mathematics books as a general formula in a mathematical notation as follows:

$$P(A) = \frac{s}{n}$$

where s = number of successful outcomes
and n = total number possible outcomes
and P(A) means, "the probability that event A will occur."

EXAMPLE PROBLEM:
What is the probability of rolling a number greater than four on a single roll of one die?

SOLUTION:
There are two numbers on a single die greater than 4, namely 5 and 6. Therefore, s = 2. There are six numbers from 1 to 6 on a single die. Therefore, n = 6.

$$P(A) = \frac{s}{n} = \frac{2}{6} = \frac{1}{3}$$

1/3 is about .333 or 33.3%. One out of three rolls should yield a number greater than four which is about 33.3% of the time. (We converted the fraction to decimal and then to percent.)

EXPERIMENT:
Verify by observation that the solution is a reasonable one. Take a single die and roll it 30 times. Record the number that is rolled each time, then count the number of times out of 30 that a number greater than four appears. Place that number in the numerator and 30 in the denominator. Your resulting fraction should be very close to $\frac{10}{30}$ or $\frac{1}{3}$

AUTHOR'S EXPERIMENT:
I conduced this experiment and rolled the following thirty numbers: 5,4,5,5,6,1,6,4,2,4,1,4,6,1,2,4,4,6,4,2,5,6,6,3,4,3, 5,6,4,3. My results were 12/30 which means 12 out of 30 rolls were greater than four. This is very close to the "a priori" prediction which is 10 out of 30 rolls. One out of three rolls should be greater than 4.

- - - - - - - - - - - - -

<u>**13.** Probability</u> in practice <u>means probability in the long run</u>. An exact value of probability can be obtained from mathematical <u>a priori</u> calculations. However, this does not guarantee that a future event will take place as predicted by mathematical calculations. This is especially true when dealing with a small set of observed cases.

.

EXAMPLES:
(a) The author's experiment in Frame 12 was off by approximately 7%.

(b) In actual experiments when tossing a coin six times, you might get five heads and one tail instead of three heads and three tails as predicted by <u>a priori mathematical calculations.</u>

(c) If you tossed a coin 100 times, you would be exercising probability <u>in the long run.</u> In this type of situation with a large number of tosses, you will discover that the distribution of heads and tails will be nearly the same because of performing a very large number of tosses. <u>A priori</u> probabilities are derived from mathematical reasoning rather than observations but they do not predict with 100% accuracy what the next toss will be.

14. The probability of failure of an event “A” is abbreviated in mathematical notation by “P (not A).”

$$P(\text{ not } A) = 1 - P(A)$$
$$= 1 - \frac{s}{n}$$

As was indicated in Frame 12, s is the total number of possible successful outcomes and n is the total number of outcomes. One represents absolute certainty of an event.

EXAMPLE PROBLEMS:

(a) What is the probability of failure in rolling a three on a single toss of one die? (Let “A” represent a roll of 3.)

SOLUTION:

$$P(\text{ not } A) = 1 - P(A)$$
$$= 1 - \frac{s}{n}$$
$$= 1 - \frac{1}{6} = - \frac{5}{6}$$

This answer means that five out of six rolls should not yield a three in the long run during many rolls of the dice.

(b) What is the probability of failure in rolling a number greater than four on a single toss of one die? (Let B represent a roll greater than four.)

SOLUTION:

$$P(\text{not } B) = 1 - P(B)$$
$$= 1 - \frac{s}{n}$$
$$= 1 - \frac{2}{6} = \frac{4}{6} = \frac{2}{3}$$

This means two out of three rolls should not yield a number greater than four in the long run of many rolls.

15. EXERCISE 3

Answer all questions before checking the answers in Frame 16 on the next page.

1. What kind of numbers are used to represent probability?

____

2. In the following formulas fill in the missing words or numbers:

(a) $P(\text{an event}) = \dfrac{\text{the number of possible (} \qquad \text{) outcomes}}{\text{(} \qquad \qquad \text{) of possible outcomes}}$

(b) P(not A) = ________________________________

3. What is the probability of rolling on a single toss of one die the following:

(a) a two ________________________________
(b) an even number________________________________
(c) a nine________________________________
(d) any number________________________________
(e) a number less than three________________________
(f) a number less than seven________________________
(g) a number evenly divisible by two________________
(h) a number different from four.

4. Find the probability of failure in each case of question 3.

5. Probability in practice does not always come out in _______ runs. In the ______ run you will come closer to the predicted probability outcomes.

16. ANSWERS FOR EXERCISE 3

1. An event that is certain to happen has a probability of one. An event that cannot take place has a probability of zero. The remaining probabilities fall between one and zero. These are fractional probabilities often expressed as percent.

2. (a) successful in the numerator; total number in the denominator

   (b) P(not A) = 1 - P(A) (or) P(not A) = $1 - \frac{s}{n}$

3. (a) $\frac{1}{6}$ (This means one out of six rolls should yield a two.)

   (b) $\frac{3}{6}$ or $\frac{1}{2}$ (This means one out of two rolls should yield an even number.)

   (c) 0 (This means the event cannot occur.)

   (d) 1 (The event is certain.)

   (e) $\frac{2}{6}$ or $\frac{1}{3}$

   (f) same answer as (d)

   (g) $\frac{3}{6}$ or $\frac{1}{2}$

   (h) $\frac{5}{6}$

4. (a) $\frac{5}{6}$ (b) $\frac{1}{2}$ (c) 1 (d) 0 (e) $\frac{2}{3}$ (f) 0 (g) $\frac{1}{2}$ (h) $\frac{1}{6}$

5. short; long

17. Fair odds require that both the player and the person running a betting game break even in the long run. However, casinos do not offer fair odds. The house enjoys an advantage in every game that the casino offers. Some games offer better odds than other games. Slots usually offer the worst odds. Smart people avoid playing the slots.

EXAMPLES OF FAIR ODDS:
(a) Two out six numbers on a single die are less than three. The two numbers (1 and 2) should show up 2 out of 6 rolls or 1 out of 3 times in the long run. The probability is 1/3 and the odds in favor of the two numbers showing up are "1 to 2." If you bet $1.00 on rolling a number less than three on a single toss of the die, then a fair return would be $2.00 each time you rolled less than three. The $1.00 bet should also be returned along with the $2.00 payoff.

(b) Here is a list fair returns on a $1.00 bet in three different cases of "odds in favor."

odds in favor	"1 to 3"	"1 to 5"	"1 to 1000"
winner's fair return	$3.00	$5.00	$1000.00
odds against	"3 to 1"	"5 to 1"	"1000 to 1"

NOTES:
(1) "Odds in favor" consist of a pair of numbers which tell us the likelihood that a particular event will take place. "Odds against" tell us the likelihood that an event will not take place. In gambling situations the "odds against" are usually stated with the exception of horse racing.
(2) The odds against rolling a 3 with a single die can be stated in a gambling situation as being 5 to 1. However, in

probability, the likelihood of rolling a 3 would be stated as 1/6 or about 16.7%.

REVIEW:

The difference between the probability formula and the odds formula is evident by examining each formula used to represent each concept. As stated earlier, the probability of an event is given by the formula notation as follows:

$$P(A) = \frac{s}{n}$$

where s = number of successful outcomes
and n = total number possible outcomes
and P(A) means, "the probability that event A will occur."

Odds are given by the following formula:

Odds = (Chances for) : (Chances against)

Let's give an example and apply each formula:

The odds of rolling a number less than 3 with a single die are 2 to 4 or 1 to 2. The probability of rolling a number less than 3 is 2/6 or 1/3.

Keep in mind that probability and odds use two different formulas to describe the likelihood that an event will take place. They offer a slightly different numerical comparison for a given gambling situation. In some published articles the two concepts are used interchangeably and this causes confusion.

18. You can roll a pair of dice in one of 36 combinations.

(1) Consider a pair of dice where one is blue and the other die is red.
(2) The sample space (i.e., a listing of all possible outcomes when rolling dice) is obtained by carrying out a cross product operation. Let B stand for the blue die and R stand for the red die. The cross product "B x R" is an operation in mathematics that produces ordered pairs of numbers.

$$B \times R = \{1, 2, 3, 4, 5, 6\} \times \{1, 2, 3, 4, 5, 6\}$$

$$B \times R = \left\{\begin{matrix} (1,1), & (2,1), & (3,1), & (4,1), & (5,1), & (6,1) \\ (1,2), & (2,2), & (3,2), & (4,2), & (5,2), & (6,2) \\ (1,3), & (2,3), & (3,3), & (4,3), & (5,3), & (6,3) \\ (1,4), & (2,4), & (3,4), & (4,4), & (5,4), & (6,4) \\ (1,5), & (2,5), & (3,5), & (4,5), & (5,5), & (6,5) \\ (1,6), & (2,6), & (3,6), & (4,6), & (5,6), & (6,6) \end{matrix}\right\}$$

NOTE: The above set contains all possible combinations when rolling a pair of dice. The first number of each pair represents the blue die and the second number represents the red die.
(3) Verify the following statements by examining the above set of ordered pairs of numbers:
(a) That there are 36 combinations. (Count them.)
(b) That out of the 36 combinations:
1. Four pairs add up to five (e.g. (3, 2) adds up to five).
2. The probability of rolling a five must be 4/36 or 1/9.
3. P(failure) is 32/36 or 8/9.
4. The odds in favor of rolling a five: "1 to 8."
5. The odds against of rolling a five : "8 to 1."
Therefore, A "fair bet" would require that the player receive $8.00 for each $1.00 bet that wins.

19. In Part (3) of Frame 18, you were asked to verify some statements. Here is an explanation for each statement:
(a) We can count 36 pairs of numbers in the cross product set. There are 36 combinations when rolling fair dice.

(b) Out of the 36 combinations:

1. Four pairs add up to five. 4 pairs out of 36 show a 5.

2. P(rolling a 5) $= \frac{s}{n} = \frac{4}{36} = \frac{1}{9}$ This means one out of nine rolls should yield a five (in the long run.)

3. P(failure) $= \frac{8}{9}$ which means eight out of nine rolls should not yield a five combination.

4. Odds in favor (rolling a 5) $= \frac{P(\text{success})}{P(\text{failure})} = \frac{\frac{1}{9}}{\frac{8}{9}}$

$$= \frac{1}{9} \div \frac{8}{9} = \frac{1}{9} \times \frac{9}{8}$$

$= \frac{1}{8}$ or "1 to 8" are the odds in favor

5. Odds against (rolling a 5) $= \frac{8}{1}$ or "8 to 1 against."

NOTE: By adding each pair of numbers in the cross product shown in Frame 18, we can show the actual sum of each pair and derive from such a chart the actual number of times a particular sum shows up. For example, the sum 7 appears 6 times. How many times does the sum 6 show up? Answer: 5 times. How many times does the sum 10 show up? Answer: 3 times. 10 shows up 3 out of 36 times so the probability of rolling a 10 is 3/36 or 1/12. In the long run a roll of 10 will occur 1 out of 12 times.

20. EXERCISE 4

Answer all questions before checking the answer key in Frame 21.

1. Fair odds require that both the player and the person running a betting game ___________ ___________ in the long run.

2. Do state lotteries offer "fair betting odds?" Explain your answer? ____

____

____

3. State the formulas for probability and odds and explain the difference between the two formulas. ___________

____

____

____

4. What are the "fair betting odds" for rolling an even number on a single toss of one die? ________________

5. What are the "fair betting odds" for rolling a number greater than four on a single toss of one die?_________

6. The sample space for rolling dice contains ______ pairs of numbers.

7. The sum of each pair of numbers in the sample space represents ___________________________________ .

8. An alternate method of showing the possible outcomes when rolling a pair of dice is a table. Examine this table and compare it with the cross product in Item (2) of Frame 18, then answer the questions below the table.

		Red die					
		1	2	3	4	5	6
Blue die	1	(1, 1),	(2, 1),	(3, 1),	(4, 1),	(5, 1),	(6, 1)
	2	(1, 2),	(2, 2),	(3, 2),	(4, 2),	(5, 2),	(6, 2)
	3	(1, 3),	(2, 3),	(3, 3),	(4, 3),	(5, 3),	(6, 3)
	4	(1, 4),	(2, 4),	(3, 4),	(4, 4),	(5, 4),	(6, 4)
	5	(1, 5),	(2, 5),	(3, 5),	(4, 5),	(5, 5),	(6, 5)
	6	(1, 6),	(2, 6),	(3, 6),	(4, 6),	(5, 6),	(6, 6)

In the following questions, assume a single toss of a pair of dice.

(a) How many pairs add up to seven? ________
(b) What is the probability of rolling a seven? ________
(c) What is the probability of not rolling a seven? ____
(d) What are the odds in favor of rolling a seven? ____
(e) What are the odds against rolling a seven? ________
(f) How many pairs add up to nine? ________
(g) What is the probability of rolling a nine? ________
(h) What is the probability of not rolling a nine? ____
(i) What are the odds in favor of rolling a nine? ____
(j) What are the odds against rolling a nine? ________

21. ANSWERS FOR EXERCISE 4

1. break even
2. States do not operate on a break even basis. They make a profit by creating unfair odds (against the betting public.)
3. $\text{Probability} = \frac{\text{desired outcomes}}{\text{total number of possible outcomes}}$ $\quad P(A) = \frac{s}{n}$

   $\text{Odds in favor} = \frac{P(\text{success})}{P(\text{failure})}$
4. Odds in favor are "1 to 1."
5. Odds in favor are "1 to 2." Odds against are "2 to 1."
6. 36
7. The roll of the dice.
8. (a) 6 (b) $\frac{6}{36}$ or $\frac{1}{6}$ (c) $\frac{30}{36}$ or $\frac{5}{6}$ (d) "1 to 5"

   (e) "5 to 1" (f) 4 (g) $\frac{4}{36}$ or $\frac{1}{9}$ (h) $\frac{32}{36}$ or $\frac{8}{9}$

   (i) "1 to 8" (j) "8 to 1"

SUMMARY OF FORMULAS

$$\text{Probability} = \frac{\text{the number of desired outcomes}}{\text{total number of possible outcomes}}$$

$$\text{Odds in favor} = \frac{P(\text{success})}{P(\text{failure})}$$

Empirical probability (E.P.) based on observations

$$E.P. = \frac{\text{number of observed heads}}{\text{the total number of outcomes}}$$

Odds = (chances for) : (chances against)

(Chance against) = (total chances) - (chances for)

$$P(A) = \frac{s}{n}$$

$$P(\text{not } A) = 1 - P(A)$$
$$= 1 - \frac{s}{n}$$

Recent titles published by William R. Parks

www.wrparks.com

Boolean Algebra and Switching Circuits

Computer Number Bases

Handbook for Piano Practice

Colours of Fire

Letters to a Young Math Teacher

Program Your Calculator

A Franciscan Odyssey

The Nature Watch Collection Book One

The Nature Watch Collection Book Two

Political Economics

Peppin Puffin to the Rescue

Choices

Time to Fly

Jonah the Reluctant Prophet

Earth, God's Garden

Made in the USA
Coppell, TX
09 December 2019